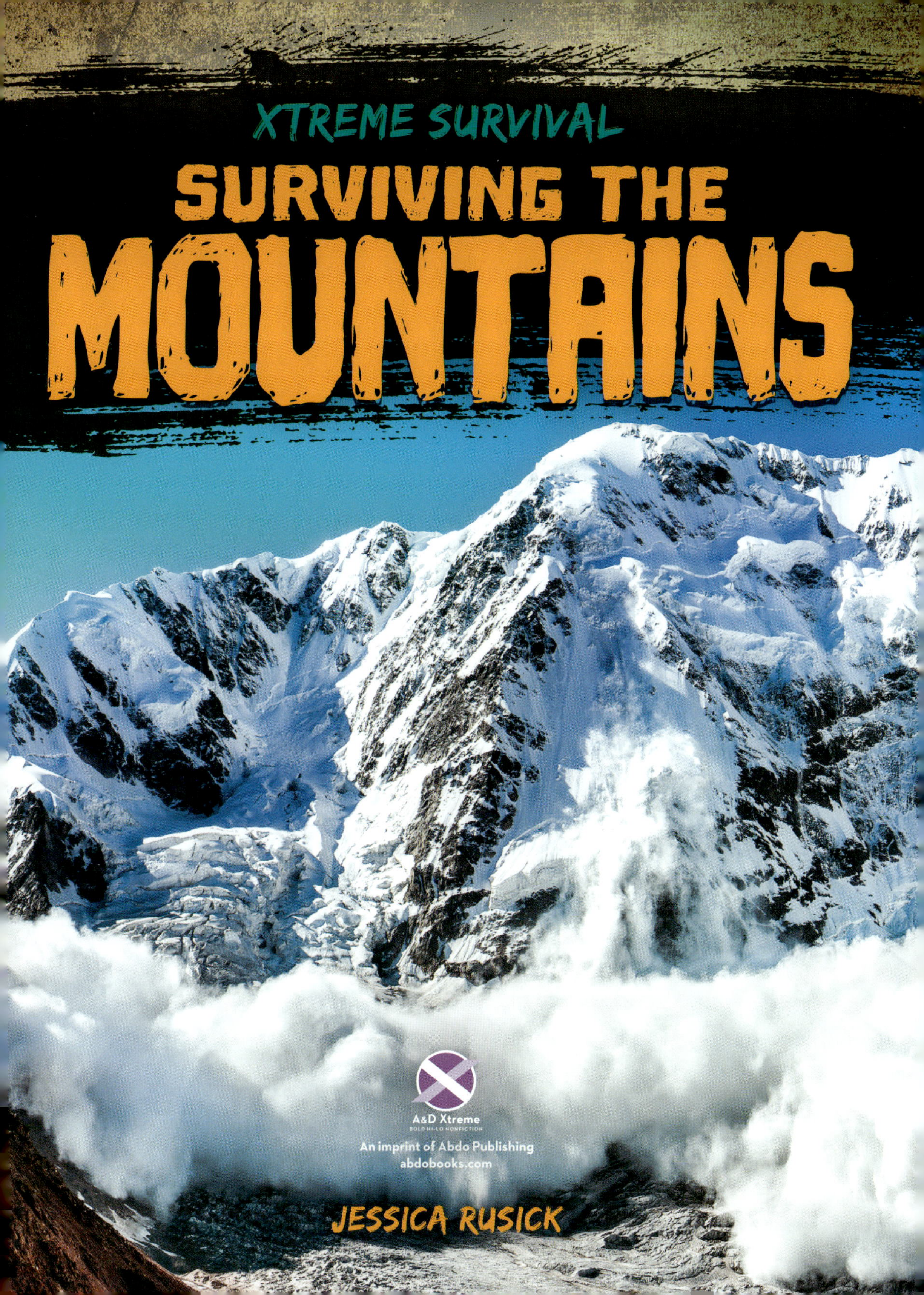
XTREME SURVIVAL
SURVIVING THE
MOUNTAINS
A&D Xtreme
BOLD HI-LO NONFICTION
An imprint of Abdo Publishing
abdobooks.com
JESSICA RUSICK

TAKE IT TO THE XTREME!

GET READY FOR AN EXTREME ADVENTURE! THE PAGES OF THIS BOOK WILL TAKE YOU INTO THE THRILLING WORLD OF OUTDOOR SURVIVAL. WHEN YOU HAVE FINISHED READING THIS BOOK, TAKE THE XTREME CHALLENGE ON PAGE 45 ABOUT WHAT YOU'VE LEARNED!

ABDOBOOKS.COM
Published by Abdo Publishing, a division of ABDO, PO Box 398166, Minneapolis, Minnesota 55439.

A&D Xtreme™ is a trademark and logo of Abdo Publishing.
Printed in the United States of America, North Mankato, MN.
102023
012024

Design: Tamara JM Peterson, Mighty Media, Inc.
Editor: Katherine Chu
Cover Photograph: Lysogor Roman/Shutterstock Images
Interior Photographs: Abramov Michael/Shutterstock Images, pp. 28-29; Arand/iStockphoto, pp. 38-39; Arte Index/iStockphoto, pp. 32-33; Avatar_023/Shutterstock Images, pp. 18-19; BlueBarronPhoto/Shutterstock Images, pp. 4-5; eivanov/Shutterstock Images, pp. 12-13; Ekspansio/iStockphoto, pp. 24-25; EvgeniyQ/iStockphoto, pp. 34-35; Jaybirdphotography/iStockphoto, pp. 16-17; Jonathan Fox/Wikimedia Commons, pp. 6-7; kasakphoto/Shutterstock Images, pp. 8-9; kovop/Shutterstock Images, p. 44; LifestyleVisuals/iStockphoto, pp. 14-15;Lysogor Roman/Shutterstock Images, p. 1; Marcus Placidus/Shutterstock Images, p. 46; mikolajn/iStockphoto, pp. 42-43; nyker/Shutterstock Images, pp. 40-41; Patryk Kosmider/Shutterstock Images, pp. 36-37; Pavel Vaschenkov/Shutterstock Images, pp. 22-23; Premysl/Shutterstock Images, pp. 20-21; Tom Grundy/Alamy Photo, pp. 10-11; Uwe Moser/iStockphoto, pp. 30-31; Ville Kangas/Shutterstock Images, pp. 26-27
Design Elements: Nik Merkulov/Shutterstock Images (grunge background); queezz/Shutterstock Images (brush texture)

Library of Congress Control Number: 2023939323

Publisher's Cataloging-in-Publication Data
Names: Rusick, Jessica, author.
Title: Surviving the mountains / by Jessica Rusick
Description: Minneapolis, Minnesota : Abdo Publishing, 2024 | Series: Xtreme survival | Includes online resources and index.
Identifiers: ISBN 9781098291839 (lib. bdg.) | ISBN 9781098278731 (ebook)
Subjects: LCSH: Survival--Juvenile literature. | Survival skills--Juvenile literature. | Mountaineering accidents--Juvenile literature. | Mountaineering--Search and rescue operations--Juvenile literature. | Outdoor life--Juvenile literature. | Wilderness survival--Juvenile literature.
Classification: DDC 613.69--dc23

TABLE OF CONTENTS

CHAPTER 1

A MOUNTAIN HIKE

Kings Canyon National Park has the largest grove of sequoia trees in the world.

In July 2014, Gregory Hein planned to spend several days hiking on Mount Goddard in California's Kings Canyon National Park. Hein was an experienced hiker, but he had not hiked regularly for a few years.

Mount Goddard has a 13,568-foot (4,136 m) peak and is part of the Sierra Nevada mountain range.

Hein reached the top of Mount Goddard in two days. He started hiking off-trail to a nearby basin. But as he **descended** the rocky, snowy mountain, a boulder shifted loose under his hand. It crashed into him, breaking his leg and knocking him down the mountainside. Hein was **stranded** in the mountains with a severe injury. How would he survive?

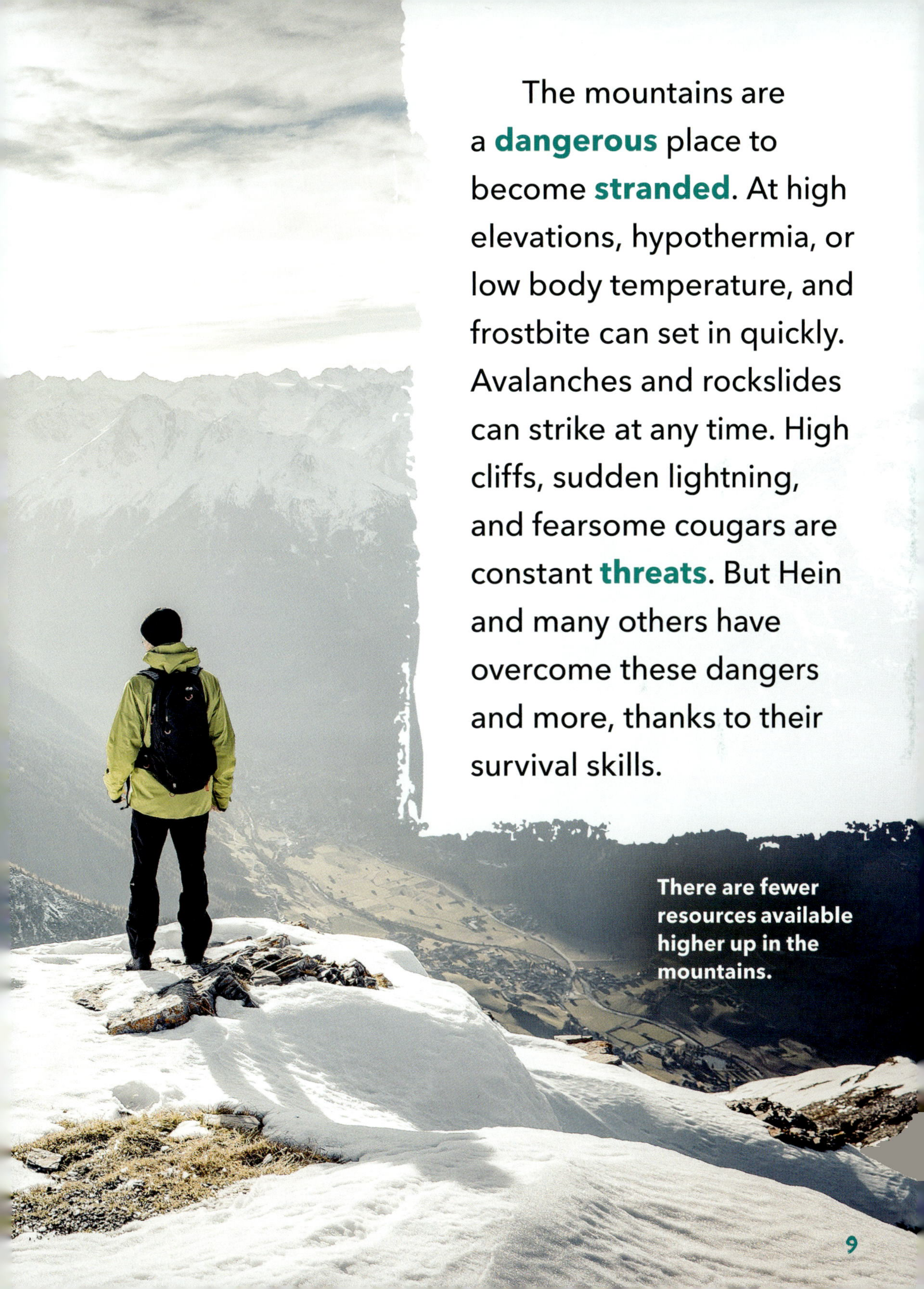

The mountains are a **dangerous** place to become **stranded**. At high elevations, hypothermia, or low body temperature, and frostbite can set in quickly. Avalanches and rockslides can strike at any time. High cliffs, sudden lightning, and fearsome cougars are constant **threats**. But Hein and many others have overcome these dangers and more, thanks to their survival skills.

There are fewer resources available higher up in the mountains.

SURVIVOR SPOTLIGHT

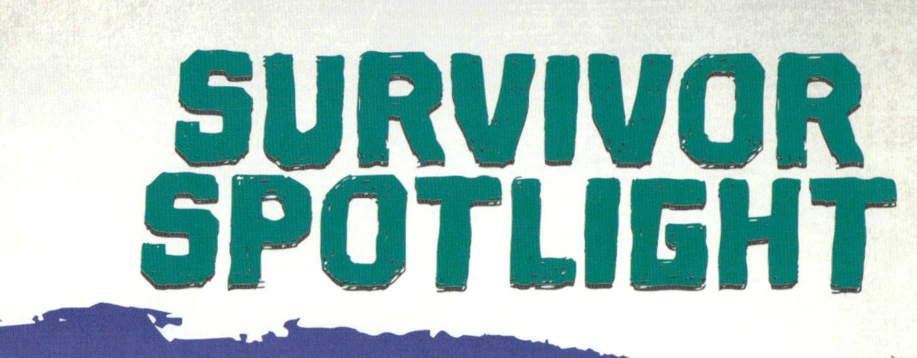

Hein used his hiking poles to make a **splint**, which supported and stabilized his leg. He cleaned his wound daily with snow and ice. Hein drank melted snow out of his hood to stay hydrated and ate crickets, moths, and ants for food. Luckily, Hein had told his father his hiking plans. This ensured that rescuers knew where to search for him. They found and rescued Hein after five days!

To get more water, Hein scooted nearly two miles (3 km) to Davis Lake.

CHAPTER 2

FINDING WATER

Water is one of the most important things to have in a survival situation. Without water, humans become dehydrated. This leads to tiredness, light-headedness, and eventually death. Luckily, there are several water sources in the mountains, including rainwater and snow.

XTREME FACT

Humans can only survive for about three days without water.

Snow can be melted into water with fire.

Walk downhill to find water, as this is where it flows. Listen for gurgling sounds. In 2017, Shuei Kato got lost hiking in the Colorado mountains. He drank from a stream to survive. Mountain streams are often naturally clear, but they may contain harmful **pathogens** that can make you sick. However, getting sick is preferable to dying of dehydration.

If possible, purify water by boiling it or using water purification tablets before drinking it. Water purification tablets contain chemicals that break down bacteria, viruses, and parasites in water.

CHAPTER 3

STAYING WARM

Temperatures in the mountains can drop at night and at high elevations, so it's important to stay warm.

Staying warm is another key to surviving the mountains. Long exposure to the cold causes hypothermia. To stay warm, Kato moved constantly, increasing his body temperature. Body temperature naturally drops during sleep. Kato only slept briefly, eating chocolate every hour to keep himself awake and avoid hypothermia.

XTREME FACT

Hypothermia can cause death in hours.

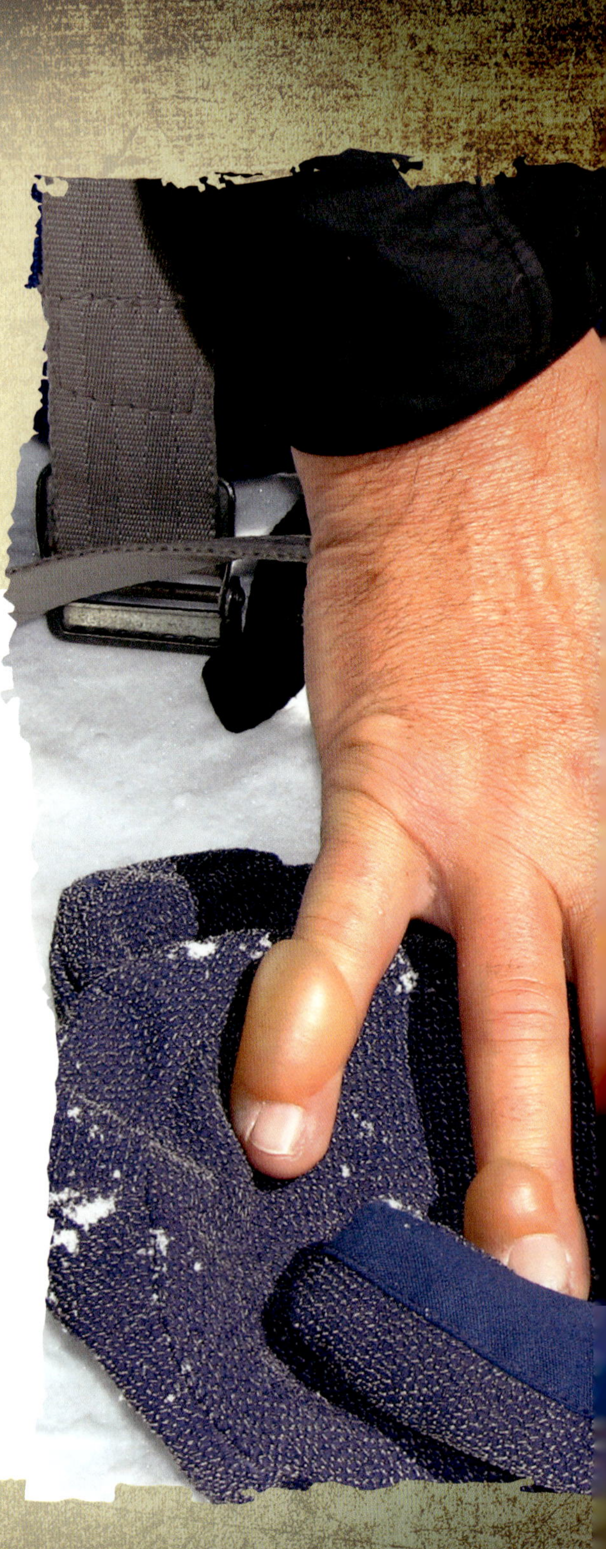

Sweat can lead to frostbite. Frostbite usually affects fingers, toes, ears, and the tip of the nose first.

Kato wore multiple layers of clothing, keeping him warm and dry. People can sweat, even in cold temperatures. Sweat can turn into frost on people's skin, making them colder. As you get hot, it's important to remove layers to avoid sweating. Then put them back on as you get cold.

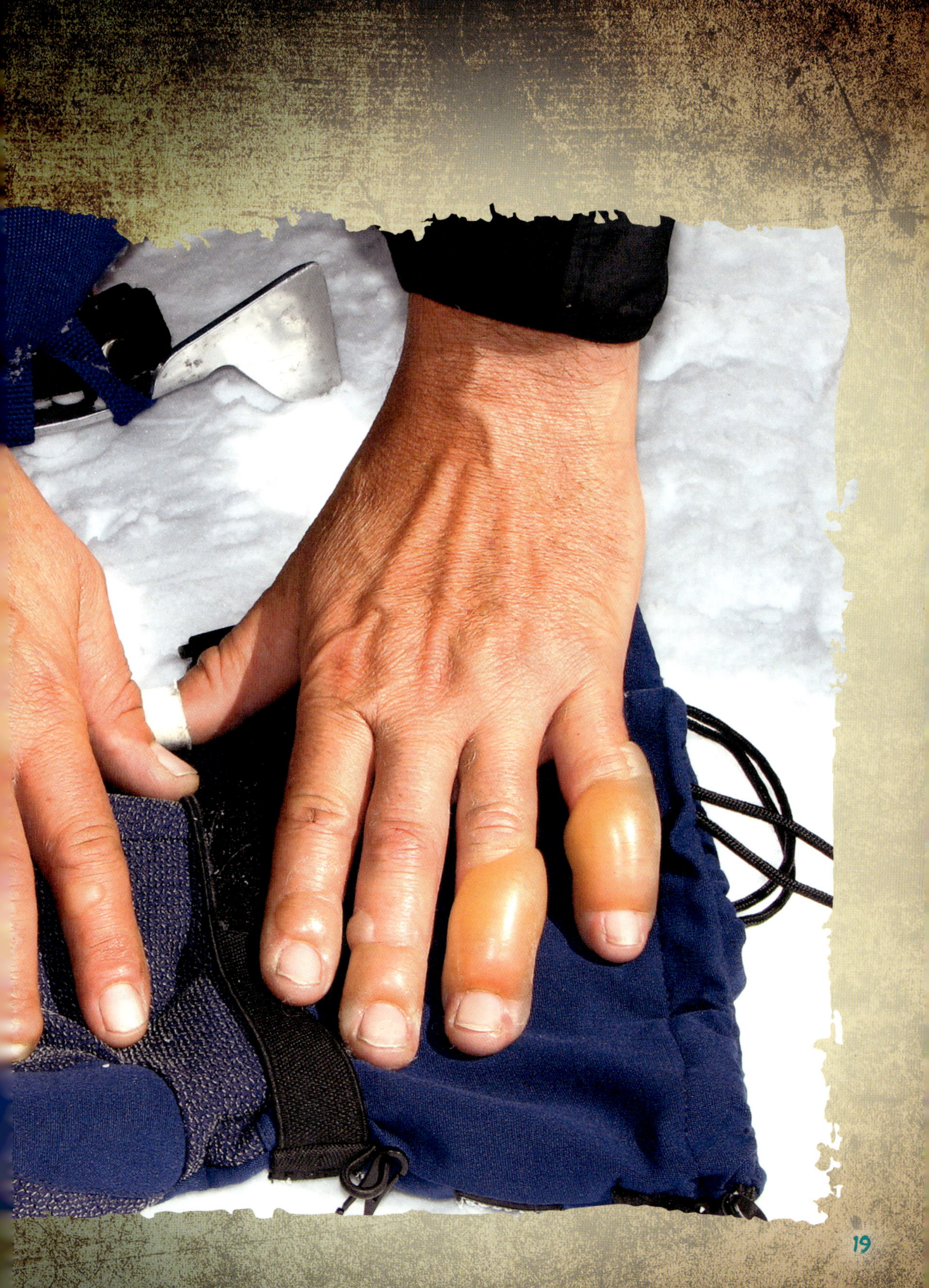

Sleeping on pine branches will help you stay dry. This helps to avoid hypothermia and frostbite.

XTREME FACT

Do not sleep directly on snow. It will make you colder!

Shelter also provides warmth. Kato slept on and under pine branches to stay warm. In 2016, Tommy Hendricks and Matthew Smith were **stranded** on a mountaintop in a snowstorm. There were few trees at the high elevation. So they sheltered under a rocky overhang and placed their feet in each other's armpits to keep warm.

Fire also keeps you warm. Use bark, twigs, pine needles, and dried grass to start the fire. Use larger sticks and logs to keep the fire burning.

Snow caves are another shelter. Dig up into snow. Create a shelf for sleeping that is higher than the entrance. This helps warm air stay in while keeping cold air out.

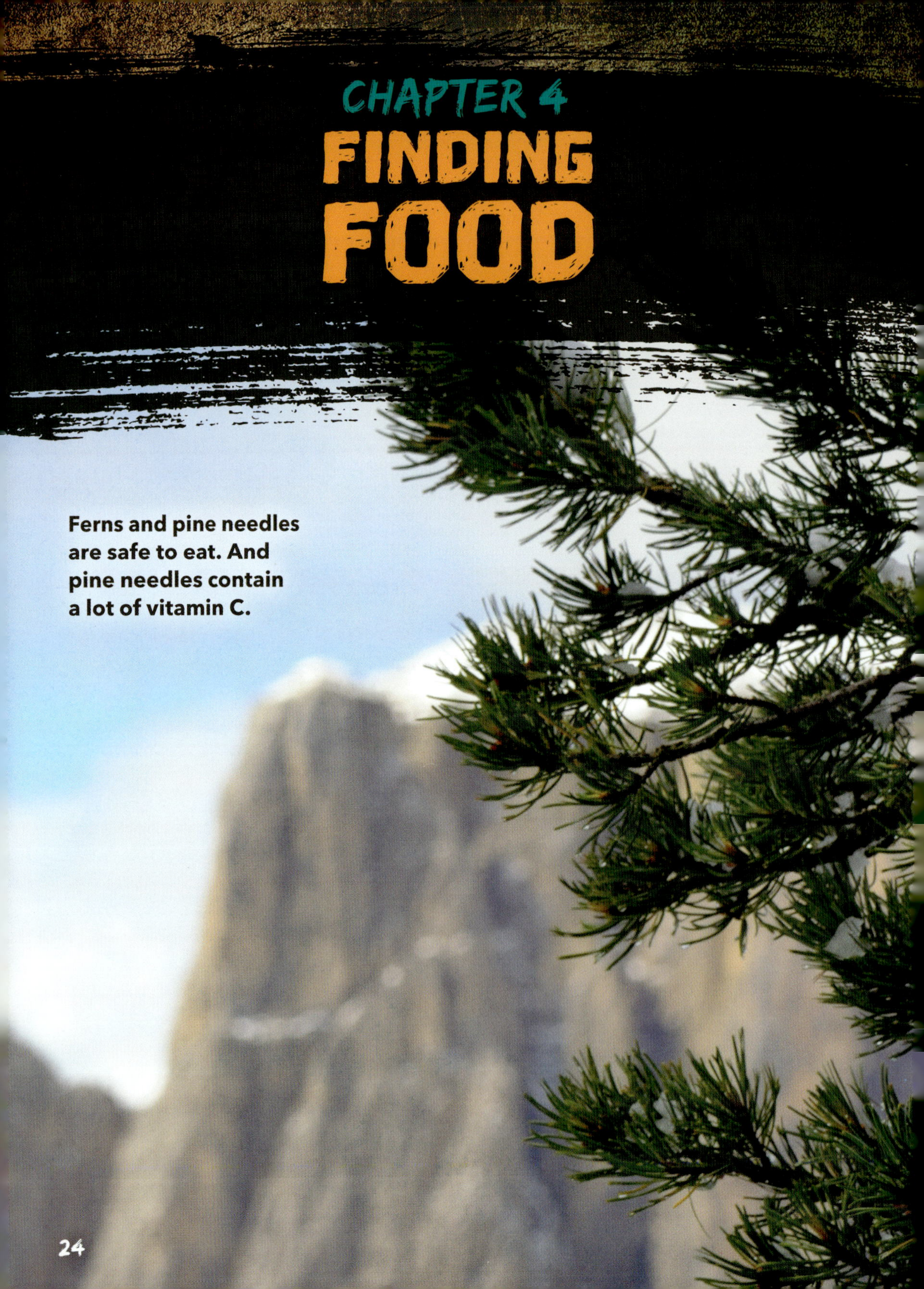

CHAPTER 4

FINDING FOOD

Ferns and pine needles are safe to eat. And pine needles contain a lot of vitamin C.

Humans can survive for weeks without food. But hunger can make it hard to **focus** and think clearly. Finding food can make survival easier. At lower elevations, there are several safe plants, bugs, and fish to eat.

XTREME FACT

Cabe made moss beds inside logs to sleep in while he was stranded.

In 2019, hiker Marshall Cabe became lost in the Washington mountains. He ate huckleberries and ants to survive. But not all plants are safe to eat. Avoid red and white plants, plants with spines, and plants you don't recognize.

Several varieties of berries and mushrooms are poisonous to humans. The destroying angel mushroom looks similar to some edible mushrooms. But eating it can result in death.

Bugs and fish can be caught by hand. To catch a fish, place your arm in deep, slow-moving water. Let your arm cool for several minutes. Then wiggle a finger to attract a fish. When the fish approaches, grab it by its gills.

If possible, cook the fish before eating it.

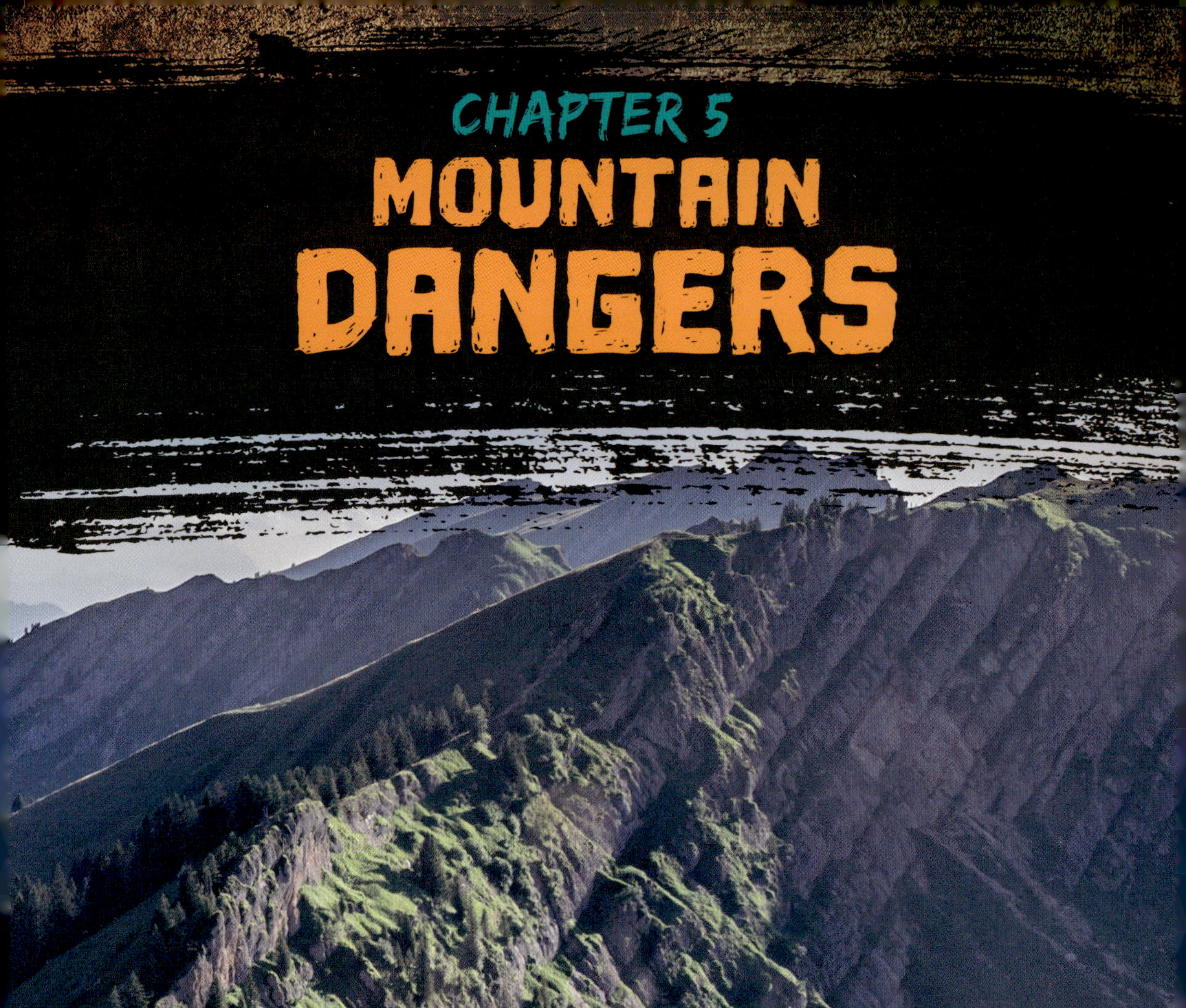

CHAPTER 5

MOUNTAIN DANGERS

At high elevations, **altitude sickness** is a danger that can cause headaches, dizziness, and shortness of breath. If untreated, it can lead to brain swelling and **hallucinations**. So it's important to **monitor** yourself for signs of altitude sickness.

The best treatments for altitude sickness include stopping to rest and moving to a lower elevation.

Avalanches are another mountain danger. They happen when snow collapses and travels quickly downhill. Heavy wind, rain, or even a person's weight can cause an avalanche. Try to avoid steep, treeless slopes of blown snow. Avalanches are more likely to happen here. Don't try to outrun an avalanche. Instead, move as far to one side as possible.

If you are buried in an avalanche, try to make an air pocket in front of your face.

Cougars are one of the most **dangerous** animals in the mountains. They may attack if startled or **provoked**. While hiking in Utah's mountains in 2020, Kyle Burgess encountered a cougar. Instead of running away, he slowly backed up while making lots of noise. The cougar followed Burgess until he found a rock to throw at it, scaring it away!

Cougars are also known as pumas, mountain lions, catamounts, or panthers. They are native to the US.

Lightning is another common danger in the mountains. Thunder always follows lightning. If you hear thunder, a storm may be approaching. Leave high areas and find shelter lower down the mountain. Avoid tall trees and water, as they attract lightning.

If there is a lightning storm, avoid lying on the ground, as it conducts electricity.

CHAPTER 6
GETTING RESCUED

One of the best ways to survive is to be prepared. Tell someone where you are going and when you expect to be back. That way, rescuers will know where to search if you get lost. Bring basic supplies such as food, water, a first aid kit, a knife, and a map and compass with you.

Even for a short hike, pack extra supplies, such as a small mirror, sleeping bag, lighter, and water purifier.

XTREME FACT

If it's safe, try to move down the mountain. There are more likely to be people, trails, and supplies there. Follow any roads or trails you find.

Kato got lost hiking up the 14,000-foot (4,267 m) Missouri Mountain in Chaffee County, Colorado.

Being lost in the mountains is scary, even if you are prepared. But it's important to stay calm. Kato **focused** on problem-solving. This helped him make smart decisions and avoid sadness and **despair**.

Many search and rescue teams are made up of volunteers.

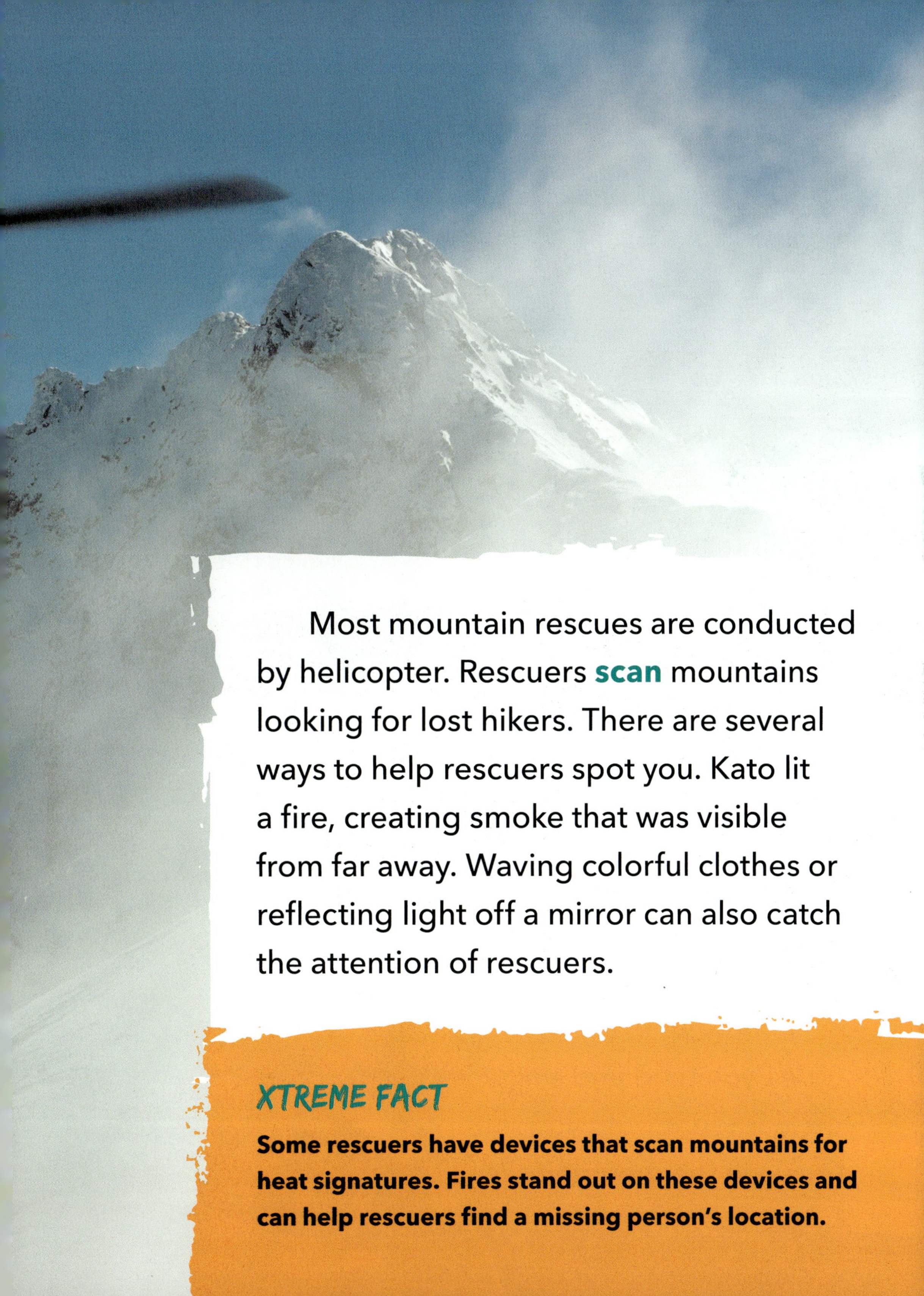

Most mountain rescues are conducted by helicopter. Rescuers **scan** mountains looking for lost hikers. There are several ways to help rescuers spot you. Kato lit a fire, creating smoke that was visible from far away. Waving colorful clothes or reflecting light off a mirror can also catch the attention of rescuers.

XTREME FACT

Some rescuers have devices that scan mountains for heat signatures. Fires stand out on these devices and can help rescuers find a missing person's location.

CHAPTER 7

SURVIVING THE MOUNTAINS

Being **stranded** in the mountains means having little **access** to food, shelter, and warmth. Lightning and avalanches are constant dangers. But with top-notch survival skills and a bit of luck, some people have overcome these challenges to survive. Think back on the mountain survival stories you read and skills you learned. Could you survive in the mountains?

It's always good to be prepared when hiking in the mountains. Wear colorful clothing so it's easier for rescue teams to spot you!

XTREME CHALLENGE

TAKE THE QUIZ BELOW AND PUT WHAT YOU'VE LEARNED TO THE TEST!

1) How did Gregory Hein care for his broken leg?

2) Why is it important for hikers to tell someone where they are going and when they expect to be back?

3) How would you stay calm if you were lost in the mountains?

4) Why did Shuei Kato only sleep for brief periods of time?

5) Would you eat bugs to survive if they were the only food source available?

GLOSSARY

access–the opportunity to gain or use something.

altitude sickness–a sickness caused by ascending too rapidly, which doesn't allow the body enough time to adjust to reduced oxygen and changes in air pressure.

dangerous–able or likely to cause hurt or harm.

descend–to go down.

despair–a belief that something good will never happen or that a problem will never be solved.

focus–to concentrate on or pay particular attention to.

hallucination–an experience in which a person senses something that isn't actually there, such as a sight or sound.

monitor–to watch, keep track of, or oversee.

pathogen—a germ that causes disease.

provoke—to cause a person or animal to become angry or violent.

scan—to examine or look over carefully.

splint—a piece of hard material used to hold a broken bone in place.

stranded—lacking the means to leave a place.

threat—something that could be harmful.

ONLINE RESOURCES

To learn more about mountain survival, please visit **abdobooklinks.com** or scan this QR code. These links are routinely monitored and updated to provide the most current information available.

INDEX